I0484371

Field-Data Collecting Form

(For Water Wells Drilling & Pumping Tests)

Well Name_____ (#_____)

Driller Company_____

Month: _____ **Year:** _____ **Place:**_____

Data collected by:_____

Address of Driller Company:_____
E-mail :_____

Φορμ Πρεπαρεδ βψ Μελεσε Γελετα Γυρμυ

Table of Contents

List of Tables and Figures

List of tables

List of figures

No figures entries found.

List of Annexes

1. **Signed Well Acceptance Letter**
2. **Signed Variation Order or Orders (If Any)**

1. Introduction

1.1 General

The well is designated by a code as _____. The contractual target depth is _____m whereas the actual drilled depth is _____m and effective drilled depth is _____m.

1.2 Location & Accessibility

The borehole site (project area) is located in/at _____ (province), _____ (country). Geographically, it is situated at a coordinate point of latitude (y) _____ and longitude (x) _____. Elevation (z) of the area is _____m above mean sea level.

Accessibility:_____
_____
_____
_____
_____

1.3 Local Geology & Hydrogeology

1.3.1 Local Geology

_____
_____
_____
_____
_____
_____
_____
_____
_____
_____

1.3.2 Hydrogeology

_____
_____
_____
_____

_____
_____
_____
_____
_____
_____
_____
_____
_____
_____
_____

2   Drilling History

2.1   Mobilization & Demobilization

Mobilization of a drilling machine (rig), assigned drilling crew and other vehicles to the drilling site took place on _____. The drilling activity commenced on _____ and completed on _____ Demobilization from the site took place on _____.

2.2   Assigned Drilling Machine & Personnel

The drilling is conducted by (Rig type):_____ drilling machine (rig) with its all the necessary equipments as well as accessories. _____ truck(s) for the supply of drilling water & fuel and _____ light vehicles are assigned. The entire drilling and pumping test works are accomplished by (driller company) _____.

Take a photo during drilling if you like

Additional comment (if any)_____
_____
_____
_____
_____
_____
_____
_____

2.3 Supervisor/s

Rank _____   Name_____

Assigned by: _____. He/ She has supervised each and every of the drilling and pumping test activities carried out on site.

2.4 Drilling Method & Activity

The upper _____m, which comprises relatively soft formation, is drilled by _____ inches (Ø) _____ (drill bit type) and _____drilling method.

The lower part, from _____m to a depth of _____m is drilled by _____ inches (Ø) _____ (drill bit type and _____drilling method. _____ used as drilling fluids.

Again the lower part, from _____m to a depth of _____m is drilled by _____ inches (Ø) _____ (drill bit type and _____drilling method. _____ used as drilling fluids.

Additional comment (if any) _____
_____
_____
_____
_____
_____
_____

Table 1: Daily drilling activities

Date	Description of the work	Remark

Date	Description of the work	Remark

Date	Description of the work	Remark

2.5 Aquifer Type and Water Strike Depths

The type of aquifer is _____. The water strike depths are inferred to be from _____m to_____ m, _____m to _____m, _____.

2.6 Geological Well Logging

Lithological sampling carried out in _____m intervals and when formation changes, during the drilling activity. The total number of samples collected is_____. Table 2 below describes the collected cutting samples along with their respective depths.

Table 2 Lithologic analysis and description

No.	Depth Range (m)		Geological description
	from	to	
1			
2			
3			

No.	Depth Range (m)		Geological description
	from	to	
4			
5			
6			
7			
8			
9			
10			
11			
12			
13			
14			
15			
16			
17			

No.	Depth Range (m)		Geological description
	from	to	
18			
19			
20			
21			
22			
23			
24			
25			
26			
27			
28			
29			
30			
31			
32			
33			
34			
35			
36			

No.	Depth Range (m)		Geological description
	from	to	
37			
38			
39			
40			

2.7   Drillers Drilling (Penetration) Rate

Table 3 Penetration rate

No.	Depth range (m)		Time Elapsed (min)	Penetration rate (m/min)
	from	to		
1	1	2		
2	2	3		
3	3	4		
4	4	5		
5	5	6		
6	6	7		
7	7	8		
8	8	9		
9	9	10		
10	10	11		
11	11	12		
12	12	13		
13	13	14		
14	14	15		
15	15	16		
16	16	17		
17	17	18		
18	18	19		
19	19	20		
20	20	21		
21	21	22		
22	22	23		
23	23	24		
24	24	25		
25	25	26		
26	26	27		
27	27	28		
28	28	29		
29	29	30		
30	30	31		

No.	Depth range (m)		Time Elapsed	Penetration rate (m/min)
	from	to	(min)	
31	31	32		
32	32	33		
33	33	34		
34	34	35		
35	35	36		
36	36	37		
37	37	38		
38	38	39		
39	39	40		
40	40	41		
41	41	42		
42	42	43		
43	43	44		
44	44	45		
45	45	46		
46	46	47		
47	47	48		
48	48	49		
49	49	50		
50	50	51		
51	51	52		
52	52	53		
53	53	54		
54	54	55		
55	55	56		
56	56	57		
57	57	58		
58	58	59		
59	59	60		
60	60	61		
61	61	62		
62	62	63		
63	63	64		
64	64	65		
65	65	66		
66	66	67		
67	67	68		
68	68	69		
69	69	70		
70	70	71		
71	71	72		
72	72	73		
73	73	74		
74	74	75		
75	75	76		
76	76	77		

No.	Depth range (m)		Time Elapsed (min)	Penetration rate (m/min)
	from	to		
77	77	78		
78	78	79		
79	79	80		
80	80	81		
81	81	82		
82	82	83		
83	83	84		
84	84	85		
85	85	86		
86	86	87		
87	87	88		
88	88	89		
89	89	90		
90	90	91		
91	91	92		
92	92	93		
93	93	94		
94	94	95		
95	95	96		
96	96	97		
97	97	98		
98	98	99		
99	99	100		
100	100	101		
101	101	102		
102	102	103		
103	103	104		
104	104	105		
105	105	106		
106	106	107		
107	107	108		
108	108	109		
109	109	110		
110	110	111		
111	111	112		
112	112	113		
113	113	114		
114	114	115		
115	115	116		
116	116	117		
117	117	118		
118	118	119		
119	119	120		
120	120	121		
121	121	122		
122	122	123		

No.	Depth range (m)		Time Elapsed	Penetration rate (m/min)
	from	to	(min)	
123	123	124		
124	124	125		
125	125	126		
126	126	127		
127	127	128		
128	128	129		
129	129	130		
130	130	131		
131	131	132		
132	132	133		
133	133	134		
134	134	135		
135	135	136		
136	136	137		
137	137	138		
138	138	139		
139	139	140		
140	140	141		
141	141	142		
142	142	143		
143	143	144		
144	144	145		
145	145	146		
146	146	147		
147	147	148		
148	148	149		
149	149	150		
150	150	151		
151	151	152		
152	152	153		
153	153	154		
154	154	155		
155	155	156		
156	156	157		
157	157	158		
158	158	159		
159	159	160		
160	160	161		
161	161	162		
162	162	163		
163	163	164		
164	164	165		
165	165	166		
166	166	167		
167	167	168		
168	168	169		

No.	Depth range (m)		Time Elapsed	Penetration rate (m/min)
	from	to	(min)	
169	169	170		
170	170	171		
171	171	172		
172	172	173		
173	173	174		
174	174	175		
175	175	176		
176	176	177		
177	177	178		
178	178	179		
179	179	180		
180	180	181		
181	181	182		
182	182	183		
183	183	184		
184	184	185		
185	185	186		
186	186	187		
187	187	188		
188	188	189		
189	189	190		
190	190	191		
191	191	192		
192	192	193		
193	193	194		
194	194	195		
195	195	196		
196	196	197		
197	197	198		
198	198	199		
199	199	200		
200	200	201		
201	201	202		
202	202	203		
203	203	204		
204	204	205		
205	205	206		
206	206	207		
207	207	208		
208	208	209		
209	209	210		
210	210	211		
211	211	212		
212	212	213		
213	213	214		
214	214	215		

No.	Depth range (m)		Time Elapsed	Penetration rate (m/min)
	from	to	(min)	
215	215	216		
216	216	217		
217	217	218		
218	218	219		
219	219	220		
220	220	221		
221	221	222		
222	222	223		
223	223	224		
224	224	225		
225	225	226		
226	226	227		
227	227	228		
228	228	229		
229	229	230		
230	230	231		
231	231	232		
232	232	233		
233	233	234		
234	234	235		
235	235	236		
236	236	237		
237	237	238		
238	238	239		
239	239	240		
240	240	241		
241	241	242		
242	242	243		
243	243	244		
244	244	245		
245	245	246		
246	246	247		
247	247	248		
248	248	249		
249	249	250		
250	250	251		

2.8 Electrical Resistivity Well Logging

Take a photo during Resistivity measurement if you like

Conducted by: _____ (company or person's name)

Date: _____, SWL_____ (m).

Table 4 Resistivity Raw Data (measured from bottom upward)

Depth (m)	Resistance (Ohm)		Resistivity (Ohm-m)	
	Short Normal (16N)	Long Normal (64N)	Short Normal (16N)	Long Normal (64N)

Depth (m)	Resistance (Ohm)		Resistivity (Ohm-m)	
	Short Normal (16N)	Long Normal (64N)	Short Normal (16N)	Long Normal (64N)

Depth (m)	Resistance (Ohm)		Resistivity (Ohm-m)	
	Short Normal (16N)	Long Normal (64N)	Short Normal (16N)	Long Normal (64N)

3   Well Construction

3.1   Installation of Surface & Production Casing

Surface casing with diameter of _____ inches is _____ (permanently or temporarily) installed for the depth of _____m. Blind and slotted production casings of _____ inches in diameter are permanently installed in the well. Production casing type_____(PVC or Steel).

Table 5: Arrangement of installed casings (bottom-top)

No.	Depth(m)		Casing arrangement	Length (m)	Remark
	From	To			
1					
2					
3					
4					
5					
6					
7					
8					
9					
10					
11					
12					
13					
14					
15					
16					
17					
18					
19					
20					
21					
22					
23					
24					
25					

Production casing stick up = _____m (above ground level)

Total length of blind casing = _____m,

Total length of Screen casing = _____m

Total length of installed casings = _____m,

3.2　Installation of Observation Pipe

Observation pipe of _____ inches in diameter and length of _____m is installed between the production casing and the well wall (annular space). The bottom _____m has been slotted. It will be serving for water level monitoring.

3.3　Gravel Packing

After installation of casing, the annular space between the installed casing and the well wall has been packed with _____ m^3 rounded and selected gravels of Ø=6-9mm.

3.4　Well Cleaning and Development

Well cleaning and development, for duration of _____ hrs, executed by using a higher capacity compressed air i.e. by _____ method to clean and improve the performance of the well in general.

Take a photo during well development if you like

3.5　Grouting & Well-head Construction

According to the agreement, after well development and cleaning, _____m top part of the well grouted by appropriate (sand, cement and crushed gravel) mixing ratio that required for mass concrete. The well-head is also constructed as per the standard design. (Yes/No)_____.

Sketch the well-head design or dimensions:

3.6 Well Capping

After finalizing the pumping test, the well has been capped with a steel plate of 6mm thickness that welded on the top of the steel production casing to prevent illegal access to the well. (Yes/No)_____

Take a photo after grouting and well head construction

4   Pumping & Recovery Tests

Is pumping test conducted? (Yes/No) _____

Conducted by: _____Date _____

4.1   Equipments Supplied

- Generator: Type: _____, Capacity =_____ KVA,
- Pump; Type: _____; Capacity H= _____m, Q=_____ l/s; HP= _____ KW ,
- Water level meter (Dip-meter) & temperature gauge and stop watch,
- For discharge measurement, 90° V-notch & 10 litre bucket ,
- Appropriate switch board, Type_____
- Riser and discharge pipes (G.S.P), Size=_____ (ø); length _____m
- Observation pipes (G.S.P), Size (ø) =_____ inch; length=_____ m,
- Crane for mantling or dismantling of pump and pipes,

4.2   Pumping Tests

Pumping test activities carried out from date_____ up to _____.

4.2.1   Provisional Test

The pump was positioned at depth of _____m below the datum point, which is _____m above ground level (top of production casing). The initial static water level is _____m. To assess the yield or capacity of the drilled well, _____ hours/min preliminary pumping with full capacity of the lowered pump (_____ l/sec discharge) was executed.

Table 6 Provisional Test Raw Data

Test Conducted by: _____	Location: _____E(X)
Borehole name: _____	_____N(Y)
Well Effective depth: _____m	Generator Capacity: _____ kVA
SWL : _____m Pump position: ____m	Discharge measuring equipment:_____
Pump Power (KW): ____, Head (H)____m	Water level meas. equipment: _____
Pump Discharge @ H (l/s): _____ l/s	Raising Main diameter: _____ inches
Testing started: Date _____, Time _____ A.M	**Test Type:** Provisional test

Time	Water Level	TDD	Discharge	Remark	Time	Water Level	RDD	Recovery	Remark
(min)	**(m)**	**(m)**	**(l/sec)**		**(min)**	**(m)**	**(m)**	**(%)**	
		Pumping test					**Recovery test**		
0					0				
1					1				
2					2				
3					3				
4					4				
5					5				
6					6				
8					8				
10					10				
12					12				
14					14				
16					16				
18					18				
20					20				
25					25				
30					30				
35					35				
40					40				
45					45				
50					50				
55					55				
60					60				
70					70				
80					80				
90					90				
100					100				
120					120				
140					140				
160					160				
180					180				
210					210				
240					240				
270					270				
300					300				
360					360				

4.2.2   Step Drawdown Test

After the provisional test and the recovery of the water level to the initial static water level four steps drawdown tests are conducted with a rate of _____, _____, _____ and _____ l/sec for duration of _____ hrs per each step.

Table 7 Step Drawdown Test Raw Data Form

Test Conducted by:_____	Location: _____E(X)			
Borehole name:_____	_____N(Y)			
Well Effective depth: _____m	Generator Capacity: _____ kVA			
SWL : _____m Pump position: ____m	Discharge measuring equipment:_____			
Pump Power (KW): ____, Head (H)_____m	Water level meas. equipment: _____			
Pump Discharge @ H (l/s): _____ l/s	Raising Main diameter: _____ inches			
Testing started: Date _____, Time _____	**Test Type:** Constant Discharge test			

Time	Water	TDD	Discharge	Remark	Time	Water	TDD	Discharge	Remark
(min)	(m)	(m)	(l/sec)		(min)	(m)	(m)	(l/sec)	
Step 1					**Step 2**				
0					0				
1					1				
2					2				
3					3				
4					4				
5					5				
6					6				
8					8				
10					10				
12					12				
14					14				
16					16				
18					18				
20					20				
25					25				
30					30				
35					35				
40					40				
45					45				
50					50				
55					55				
60					60				
70					70				
80					80				
90					90				
100					100				
120					120				
140					140				
160					160				
180					180				

Test Conducted by:_____	Location: _____E(X)
Borehole name:_____	_____N(Y)
Well Effective depth: _____m	Generator Capacity: _____ kVA
SWL : _____m Pump position: _____m	Discharge measuring equipment:_____
Pump Power (KW): ____, Head (H)_____m	Water level meas. equipment: _____
Pump Discharge @ H (l/s): _____ l/s	Raising Main diameter: _____ inches
Testing started: Date _____, Time _____	**Test Type:** Constant Discharge test

	Step 3				Step 4		
0				0			
1				1			
2				2			
3				3			
4				4			
5				5			
6				6			
8				8			
10				10			
12				12			
14				14			
16				16			
18				18			
20				20			
25				25			
30				30			
35				35			
40				40			
45				45			
50				50			
55				55			
60				60			
70				70			
80				80			
90				90			
100				100			
120				120			
140				140			
160				160			
180				180			

4.2.3   Constant Discharge Test

After the provisional test [and step drawdown test] had been conducted, and the well had recovered to acceptable static water level, constant discharge test carried out without interruption for a total duration of _____ hours with selected constant pumping rate of _____ l/sec. Just before the beginning of constant discharge test, the measured static water level was _____m and the set pump position during the test was _____m, below the reference point. The total drawdown (TDD) measured for the range of the pumping time is _____m.

Table 8: Constant Discharge Test Raw Data Form

Test Conducted by: _____	Location: _____E(X)
Borehole name: _____	_____N(Y)
Well Effective depth: _____m	Generator Capacity: _____ kVA
SWL : _____m Pump position: ____m	Discharge measuring equipment:_____
Pump Power (KW): ____, Head (H)____m	Water level meas. equipment: _____
Pump Discharge @ H (l/s): _____ l/s	Raising Main diameter: _____ inches
Testing started: Date _____, Time _____A.M	**Test Type:** Constant Discharge test

Time (min)	Water level (m)	Draw Down (m)	Pumping Rate (l/s)	Remark (physical character, EC/TDS, pH, T)
0				Turbidity_____, Odour_____
1				
2				
3				
4				
5				
6				
8				
10				
12				
14				
16				
18				
20				
25				
30				
35				
40				
45				
50				
55				
60				EC = _____µS/cm, PH = _____, T = _____ ^{o}C
70				
80				
90				
100				
120				
140				
160				
180				EC = _____µS/cm, PH = _____, T = _____ ^{o}C
210				
240				
270				
300				
360				EC = _____µS/cm, PH = _____, T = _____ ^{o}C
420				
480				

Test Conducted by: _____	Location: _____E(X)
Borehole name: _____	_____N(Y)
Well Effective depth: _____m	Generator Capacity: _____ kVA
SWL : _____m Pump position: _____m	Discharge measuring equipment:_____
Pump Power (KW): ____, Head (H)_____m	Water level meas. equipment: _____
Pump Discharge @ H (l/s): _____ l/s	Raising Main diameter: _____ inches
Testing started: Date _____, Time _____A.M	**Test Type:** <u>Constant Discharge test</u>

Time (min)	Water level (m)	Draw Down (m)	Pumping Rate (l/s)	Remark (physical character, EC/TDS, pH, T)
540				
600				
660				
720				EC = _____ µS/cm, PH = _____, T = _____ °C
780				
840				
900				
960				
1020				
1080				
1140				
1200				
1260				
1320				
1380				
1440				EC = _____ µS/cm, PH = _____, T = _____ °C

4.3 Recovery Test

Recovery test is conducted after constant discharge test for the duration of _____ hours after pumping had been shut-off. Within the span of time, the total residual drawdown of the well was _____m. Within this duration, the well recovered for about _____% of the TDD.

Table 9: Recovery Test Raw Data Form

Test Conducted by: _____	Location: _____E(X)
Borehole name: _____	_____N(Y)
Well Effective depth: _____m	Generator Capacity: _____ kVA
SWL : _____m Pump position: _____m	Discharge measuring equipment:_____
Pump Power (KW): _____, Head (H)_____m	Water level meas. equipment: _____
Pump Discharge @ H (l/s): _____ l/s	Raising Main diameter: _____ inches
Testing started: Date _____, Time _____A.M	**Test Type:** Recovery test

Time (min)	Time Since pump started (min)	Water level (m)	Residual DD (m)	Pumping Rate (l/s)	Remark
0					
1					
2					
3					
4					
5					
6					
8					
10					
12					
14					
16					
18					
20					
25					
30					
35					
40					
45					
50					
55					
60					
70					
80					
90					
100					
120					
140					
160					
180					
210					
240					
270					
300					

Test Conducted by: _____	Location: _____E(X)
Borehole name: _____	_____N(Y)
Well Effective depth: _____m	Generator Capacity: _____ kVA
SWL : _____m Pump position: _____m	Discharge measuring equipment:_____
Pump Power (KW): ____, Head (H)_____m	Water level meas. equipment: _____
Pump Discharge @ H (l/s): _____ l/s	Raising Main diameter: _____ inches
Testing started: Date _____, Time _____A.M	**Test Type:** <u>Recovery test</u>

Time (min)	Time Since pump started (min)	Water level (m)	Residual DD (m)	Pumping Rate (l/s)	Remark
360					
420					
480					
540					
600					
660					
720					

5 Water Sampling & Quality Analysis

5.1 Water Sampling

Towards the end of the constant rate-pumping test, required amount i.e._____ litter/s of water sample is taken for a laboratory (Physico-chemical) analysis.

5.2 On Site Physical Analysis

The physical inspection of the water flowing out of the well during the pumping period revealed that the water from the well is odourless, tasteless and colourless. Water sample collected in 2 litter plastic bottles at the end of pumping test.

Table 10: On site water quality test

No	Parameter	Field test result	Remarks
1	Turbidity		
2	Colour		
3	Odour		
4	pH		
5	EC		
6	Temperature (°C)		
7			
8			

6 PERMENANT PUMP & GENERATOR INSTALLATION & PIPE LAYING (if any)

Based on the initial agreement, permanent a pump and a Generator were installed for the well. Yes/No _____. Pipe-laying was carried out. Yes/No _____.

6.1 Permanent pump installed

A Pump with *control panel, automatic electrodes and raising main* were installed at the borehole. Type of Pump installed: _____, Model _____, Control panel type_____.

Pump diameter: _____ **inches**
Pump Capacity:
      Power (kW) _____
      Head (m) _____
      Discharge (l/s) @ the given head: _____ **l/sec**
Raising Main type: _____

Raising Main diameter: _____inch

Pump Position (b. g. l): _____m

Additional
comment_____

_____

_____

_____

6.2 Permanent Generator installed

Type of Generator: _____, Engine No. _____

Capacity (KVA): _____

(Take a Photo if you like)
- *Generator & Control board*

6.3 Pipe laying

_____inches pipe and different fittings & flow meter were installed in the pipes laying. Type and quantity of pipes, fittings and flow meter is shown in the table below.

Table 11 Summary of works executed and materials consumed

No	Item/activity description	Unit	Qty	Remark
1	**Drilling & Well development**			
1.1	Drilling by _____ " bit	m		
1.2	Drilling by _____ " bit	m		
1.3	Drilling by _____ " bit	m		
1.4	Well development	hrs		
2	**Materials consumed during drilling & construction (supplied by EWDU)**			
2.1	Foam	Lit		
2.2	Bentonite	Kg		
2.3	Hammer oil	Lit		
2.4	Cement for well head construction & grouting	bags		
2.5	Sand	m³		
2.6	Crushed gravel	m³		
2.7	Welding electrode 3.2mm	Packet		
2.8	Cutting disc	Pcs		
2.9	Grinding Disc	Pcs		
2.10	Insulating tape (normal)	Pcs		
2.11	high voltage insulating tape	Roll		
2.12				
2.13				
2.14				

No	Item/activity description	Unit	Qty	Remark
3	**Supply & installation of materials**			
3.1	Supply & installation of ____" _____ blind casing	m		_____m is stick up a. g. l
3.2	Supply & installation of ____" _____ Screen casing	m		
3.3	Supply & installation of _____" Permanent surface casing	m		
3.4	Supply & installation of _____" Observation pipe	m		
3.5	Supply & installation of (diameter) _____ mm size well rounded river gravel	m³		
3.6	Supply & installation of _____" Water Meter	No		
3.7	Supply & installation of ___" GS Pipe, Class_____ (for raising main and connection to distribution line	m		
3.8	Supply & installation of Gate Valve _____"	Pcs		
3.9	Supply & installation of Check Valve _____"	Pcs		
3.10	Supply & installation of Elbow _____"	Pcs		
3.11	Supply & installation of Nipples _____"	Pcs		
3.12	Supply & installation of Union _____"	Pcs		
3.13	Supply & installation of Plug _"	Pcs		
3.14	Supply & installation of reducer ____" to _____"	Pcs		
3.15	Supply & installation of Nipple_____"	Pcs		
3.16	Supply & installation of Plug ____"			
3.17	Supply & installation of Plug ___"			
3.18	Supply & installation of Generator, _____ KVA,	Set		Engine No. _____, Type_____
3.19	Pump (Type_____), Model _____, Power = _____ KW, Head = _____m, Q@H = _____ l/s, with Control panel & all accessories	Set	1	
4	**Drilling tools and equipments used during drilling, construction & testing activities**			
4.1	___" _____ bit	Pcs		
4.2	___" _____ bit	Pcs		
4.3	_____" tungsten carbide bit	Pcs		
4.4	_____" Hammer (DTH) bit	Pcs		For DTH drilling

No	Item/activity description	Unit	Qty	Remark
4.5	_____" Hammer (DTH) bit	Pcs		For DTH drilling
4.6	Installation & removal of ____" surface casing	m		Temporary surface casing
4.7	EC Meter	Set		For in situ water quality measurement
4.8	PH Meter	Set		
4.9	Dipmeter (Water level indicator)	set		
4.10	Geophysical instrument for borehole electrical logging	Set		Type: _____
4.11	Plumbing equipments (Pipe threader, pipe wrenches, etc)	set		
4.12	Welding Machine	set		
4.13	GPS (to locate sites)	set		
4.14	90° V-Notch (Bucket)	No		For well discharge measurement
5	**Fuel, Lubricant & drilling water Consumed**			
5.1	Gas oil	Lit		
5.2	Gear oil N0. 10	Kg		
5.3	Engine oil	Kg		
5.4	Grease	Kg		
5.5	Drilling water	Lit		

7   Overall Remark (if Any)

7.1   Drilling Activity

_____
_____
_____
_____
_____
_____
_____
_____
_____
_____
_____
_____
_____
_____
_____
_____
_____
_____
_____
_____
_____
_____
_____

7.2   Pumping Test Activity

_____
_____
_____
_____
_____
_____
_____
_____
_____
_____
_____
_____

Annexes

3. **Signed Well Acceptance Letter**
4. **Signed Variation Order or Orders (If Any)**

.

www.ingramcontent.com/pod-product-compliance
Lightning Source LLC
Chambersburg PA
CBHW080652180526
45168CB00008B/3402